Collins

KS2
Arithmetic
SATs 10 Minute Tests

Caroline Clissold

How to Use this Book

This book contains 25 Key Stage 2 Arithmetic tests, each designed to be completed in approximately 10 minutes.

Consisting of SATs-style questions in bite-sized chunks, each 10-minute test will help children to prepare for the SATs Arithmetic paper at home.

Clearly laid out questions and easy-to-use answers will help your child become familiar with, and gain confidence in, answering and understanding SATs-style questions.

The tests are all the same level of difficulty, which means they can be carried out in any order and at any time throughout Year 6 to provide invaluable practice for your child.

- Children should work in a quiet environment where they can complete each test undisturbed. They should complete each test in approximately 10 minutes.

- The number of marks available for each question is given at the end of the question box, with a total provided at the end of each test.

- Answers and marking guidance are provided for each test.

- A score chart can be found at the back of the book, which your child can use to record their marks and see their progress.

Acknowledgements

The author and publisher are grateful to the copyright holders for permission to use quoted materials and images.

Images are © Shutterstock.com and © HarperCollins*Publishers*

Every effort has been made to trace copyright holders and obtain their permission for the use of copyright material. The author and publisher will gladly receive information enabling them to rectify any error or omission in subsequent editions. All facts are correct at time of going to press.

Published by Collins
An imprint of HarperCollins*Publishers*
1 London Bridge Street
London SE1 9GF

HarperCollins*Publishers*
Macken House, 39/40 Mayor Street Upper, Dublin 1, DO1 C9W8, Ireland

ISBN: 9780008335885

Content first published 2019
This edition published 2020
Previously published by Letts

10 9 8 7

© HarperCollins*Publishers* Limited 2020

British Library Cataloguing in Publication Data.

A CIP record of this book is available from the British Library.

Author: Caroline Clissold
Commissioning Editors: Michelle l'Anson and Fiona McGlade
Editor and Project Manager: Katie Galloway
Cover Design: Sarah Duxbury and Kevin Robbins
Inside Concept Design: Ian Wrigley
Text Design and Layout: Jouve India Private Limited
Production: Karen Nulty
Printed in the UK by Martins the Printers

MIX
Paper | Supporting responsible forestry
FSC
www.fsc.org
FSC™ C007454

This book is produced from independently certified FSC™ paper to ensure responsible forest management.

For more information visit:
www.harpercollins.co.uk/green

Contents

10 min

1 $463 + 30 =$

1 mark

2 $25 \times 6 =$

1 mark

3 $1\frac{5}{7} - 1\frac{1}{7} =$

1 mark

4 $-6 + 12 =$

1 mark

5 $4 \times 5 \times 6 =$

1 mark

6 $563{,}365 - 199{,}999 =$

1 mark

7 $3^3 - 20 =$

1 mark

8 $5.45 \times 4 =$

1 mark

9 $\dfrac{1}{3} \div 4 =$

1 mark

10 $0.1 \times 1,000 =$

1 mark

11 15% of $540 =$

1 mark

12 $\dfrac{1}{2} \times \dfrac{3}{4} =$

1 mark

1   345 × 0 =

1 mark

2   180 ÷ 3 =

1 mark

3

Show your method

1 2 │ 3 6 0

2 marks

4   −12 − 6 =

1 mark

5 $\frac{3}{5} + \frac{3}{10} =$

1 mark

6

Show your method

3 5 │ 4 4 1 0

1 mark

7 20% of £45 =

1 mark

8

$$\begin{array}{r} 2\ 4\ 5 \\ \times \quad\ 3\ 2 \\ \hline \end{array}$$

2 marks

9 168 ÷ 5 =

1 mark

10 $\dfrac{3}{5} + \dfrac{1}{8} =$

1 mark

11 3.56 × 7 =

1 mark

12 $10^2 - 5^2 =$

1 mark

10 min

1 437 + 9 =

1 mark

2 5,609 + 1,000 =

1 mark

3 4,019 − 2,346 =

1 mark

4 21,357 + 19,985 =

1 mark

5 $\frac{2}{3} + \frac{5}{12} =$

1 mark

6 1.2 ÷ 100 =

1 mark

7 $\dfrac{1}{3} \times \dfrac{2}{3} =$

1 mark

8

$1\ 5\ \overline{)\ 2\ 2\ 5\ 0}$

Show your method

2 marks

9 $\dfrac{2}{5} \div 3 =$

1 mark

10 $21.15 \times 10 =$

1 mark

11 $275 \div 4 =$

1 mark

12 90% of $480 =$

1 mark

10 min

1 $54 = \boxed{} \times 6$

1 mark

2 $\dfrac{2}{3}$ of $42 =$

1 mark

3 $1 \times 749 =$

1 mark

4 $4,500 + 4,550 =$

1 mark

5 $\dfrac{2}{5} + \dfrac{4}{5} =$

1 mark

6 $27 \div 100 =$

1 mark

7 $6^2 \times 2 =$

1 mark

8 $2\frac{2}{3} - 1\frac{1}{6} =$

1 mark

9 $\frac{7}{8} \div 2 =$

1 mark

10 $3.25 \times 8 =$

1 mark

11 $\frac{1}{4} \times \frac{2}{5} =$

1 mark

12   10% of ☐ = 60

1 mark

10 min

1 ⬚ + 300 = 746

1 mark

2 4,568 − 2,417 =

1 mark

3 $\frac{7}{10} + \frac{3}{10} =$

1 mark

4 −25 + 30 =

1 mark

5 150,201 − 148,998 =

1 mark

6 34,258 + 27,876 =

1 mark

7 25% of 80 =

1 mark

8

$$
\begin{array}{r}
5\ 6\ 3\ 2 \\
\times \quad 2\ 5 \\
\hline
\end{array}
$$

Show your method

2 marks

9 $\dfrac{3}{5} + \dfrac{5}{6} =$

1 mark

10 $3\dfrac{1}{5} \times 3 =$

1 mark

11 $\dfrac{3}{5} \div 4 =$

1 mark

12 2.75 × 1,000 =

1 mark

Test 5 total marks ..../13

13

10 min

1 199 − ☐ = 99

1 mark

2 $\frac{3}{5}$ of 60 =

1 mark

3 180 ÷ 1 =

1 mark

4 $\frac{5}{7} + \frac{6}{7} =$

1 mark

5 549 ÷ 9 =

1 mark

6 $7^3 =$

1 mark

7 1,000,000 − 100,000 =

1 mark

8 $4\frac{1}{3} - \frac{4}{9} =$

1 mark

9 0.25 × 1,000 =

1 mark

10 80% of 150 =

1 mark

11 $\frac{3}{4} \div 5 =$

1 mark

12

Show your method

$$2\ 3\ |\ \overline{1\ 2\ 8\ 8}$$

2 marks

10 min

1 198 + 10 =

1 mark

2 $\frac{1}{8} \times \frac{1}{2} =$

1 mark

3 236 × 6 =

1 mark

4 12,576 + 199 =

1 mark

5

```
      3 6 5
  ×     1 6
```

2 marks

6 $\frac{5}{8} + \frac{7}{8} =$

1 mark

Show your method

7 $1,687 \div 7 =$

1 mark

8 $3,487 - 1,698 =$

1 mark

9 $-45 + 90 =$

1 mark

10 25% of $70 =$

1 mark

11 $\dfrac{2}{5} \div 4 =$

1 mark

12

Show your method

| 2 | 3 | 3 | 5 | 8 | 8 |

2 marks

1 254 + 6 =

1 mark

2 25 × 10 =

1 mark

3 12 × 6 =

1 mark

4 2,450 + 2,460 =

1 mark

5 $\frac{3}{5} + \frac{14}{15} =$

1 mark

6 2,100 ÷ 3 =

1 mark

7 $1\frac{2}{3} \times 6 =$

1 mark

8 $(3 + 5) \times 5 =$

1 mark

9

Show your method

$$1\ 8\ |\ 1\ 7\ 2\ 8$$

2 marks

10 30% of 270 =

1 mark

11 $\frac{2}{3} \times \frac{1}{2} =$

1 mark

12 $7.45 \times 6 =$

1 mark

10 min

1 67 + 10 =

1 mark

2 425 + 67 =

1 mark

3 3,265 + ☐ = 4,265

1 mark

4 4,356 ÷ 9 =

1 mark

5 $9^2 \times 10 =$

1 mark

6 0.34 × 3 =

1 mark

7 Show your method

$$\begin{array}{r} 3\ 2\ 5\ 6\ 7 \\ \times \quad\quad\quad 1\ 4 \\ \hline \end{array}$$

2 marks

8 $35,456 - 24,998 =$

1 mark

9 $\dfrac{3}{4} = \boxed{}\ \%$

1 mark

10 Show your method

$$1\ 2\ \overline{\left|\ 1\ 0\ 3\ 2\right.}$$

2 marks

11 $\dfrac{6}{7} - \dfrac{3}{4} =$

1 mark

12 $\dfrac{2}{5} \div 6 =$

1 mark

1 34 × 3 =

1 mark

2 72 ÷ ▢ = 12

1 mark

3 324 × 7 =

1 mark

4 54,368 − 27,476 =

1 mark

5 3,786 ÷ 6 =

1 mark

6 $\frac{5}{8} \times 5 =$

1 mark

7   12,356 + 10,444 =

1 mark

8   100 − 24 × 3 =

1 mark

9   75 ÷ 12 =

1 mark

10

Show your method

$$2\,5\,\overline{)\,2\,\,2\,\,2\,\,5}$$

2 marks

11

Show your method

```
      4 6 8 9
    ×     2 3
    ─────────
```

2 marks

12   2.564 × 100 =

1 mark

1 768 + 100 =

1 mark

2 $\frac{1}{8} + \frac{5}{8} =$

1 mark

3 3 × 7 × 5 =

1 mark

4 10,000 × 4 =

1 mark

5 250 × 5 =

1 mark

6 34 ÷ 1,000 =

1 mark

7 $\dfrac{23}{100} = \boxed{}$ %

1 mark

8 $57 - 85 =$

1 mark

9 $\dfrac{2}{5} \times \dfrac{1}{8} =$

1 mark

10 $45.35 \times 7 =$

1 mark

11 35% of $240 =$

1 mark

12

$8\ 4\ \boxed{2\ 0\ 1\ 6}$

Show your method

2 marks

10 min

1 8 × 8 =

1 mark

2 35 × 4 =

1 mark

3 25 ÷ 1 =

1 mark

4 3,750 + 2,250 =

1 mark

5 4,575 ÷ 5 =

1 mark

6 12,684 − 10,736 =

1 mark

7

$$\begin{array}{r} 2\ 4\ 3\ 7\ 5\ 2 \\ \times2\ 3 \\ \hline \end{array}$$

Show your method

2 marks

8 $5\dfrac{3}{8} - 2\dfrac{3}{4} =$

1 mark

9 $\dfrac{3}{5} \times \dfrac{5}{8} =$

1 mark

10 $0.125 + \dfrac{1}{10} =$

1 mark

11 $\boxed{} \% = \dfrac{3}{5}$

1 mark

12

$$1\ 1\ \overline{)\ 4\ 9\ 6\ }$$

Show your method

2 marks

10 min

1 $0.75 - \dfrac{1}{2} =$

1 mark

2 503 − 100 =

1 mark

3 257 + 178 =

1 mark

4 320 ÷ 4 =

1 mark

5 4,079 + 3,000 =

1 mark

6 $4^3 = 100 -$ ☐

1 mark

Test 13

7	2,500 × 3 =	**8**	$\frac{9}{10} \div 5 =$

1 mark   1 mark

9	0.125 × 1,000 =	**10**	493 ÷ 6 =

1 mark   1 mark

11	23 + 45 ÷ 9 =	**12**	8.76 × 8 =

1 mark   1 mark

10 min

1 $45 \div 100 =$

1 mark

2 $\dfrac{3}{5} + \dfrac{4}{5} =$

1 mark

3 ☐ $+ 99 = 150$

1 mark

4 $25 \times$ ☐ $= 100$

1 mark

5 $8 \times 7 =$

1 mark

6 $0 \times 15 =$

1 mark

7 23,567 − 12,784 =

1 mark

8 Show your method

$$\begin{array}{r} 3\ 4\ 6\ 5 \\ \times\quad\ \ 1\ 9 \\ \hline \end{array}$$

2 marks

9 84% of 200 =

1 mark

10 $\dfrac{7}{8}$ of 96 =

1 mark

11 Show your method

2 6 | 1 1 9 0 8

2 marks

12 35% of 140 =

1 mark

1   $\dfrac{3}{8} - \dfrac{1}{8} =$

1 mark

2   254 + 8 =

1 mark

3   546 – 287 =

1 mark

4   12 × 5 × 4 =

1 mark

5   100,000 × 4 =

1 mark

6   34,568 + 999 =

1 mark

7 $\dfrac{3}{4} - \dfrac{3}{12} =$

1 mark

8 $\dfrac{7}{8} \times 6 =$

1 mark

9

$$\begin{array}{r} 4\ 6\ 3\ 5 \\ \times \qquad 1\ 5 \\ \hline \end{array}$$

Show your method

2 marks

10 $2\dfrac{3}{8} + 1\dfrac{7}{10} =$

1 mark

11 $2.65 \times 8 =$

1 mark

12 $\dfrac{2}{3} \times \dfrac{3}{4} =$

1 mark

10 min

1 50 × 6 =

1 mark

2 137 + 80 =

1 mark

3 42 ÷ 3 =

1 mark

4 5,462 + 4,000 =

1 mark

5 546 × 7 =

1 mark

6 $\frac{3}{4}$ ÷ 2 =

1 mark

7 4,560 × 40 =

1 mark

8 3.85 ÷ 100 =

1 mark

9 −27 + 84 =

1 mark

10

$$
\begin{array}{r}
8\ 9\ 4\ 5 \\
\times\qquad 3\ 6 \\
\hline
\end{array}
$$

Show your method

2 marks

11 $\dfrac{7}{12} \div 3 =$

1 mark

12 54% of 500 =

1 mark

1   254 − 127 =

                1 mark

2   72 ÷ 9 =

                1 mark

3   25 × 7 =

                1 mark

4   63 ÷ 1 =

                1 mark

5   6,154 + 2,936 =

                1 mark

6 $250 - 5^3 =$

                1 mark

7 $0.12 \times 1,000 =$

1 mark

8

Show your method

$$2\ 6\ |\ 2\ 2\ 1\ 0$$

2 marks

9 $71 - 6 \times 6 =$

1 mark

10 $\dfrac{3}{4} \times \dfrac{2}{5} =$

1 mark

11 $1.234 \times 1,000 =$

1 mark

12 $784 \div 5 =$

1 mark

1 $195 - 100 =$

1 mark

2 $235 + 75 =$

1 mark

3 ⬜ $\div 9 = 9$

1 mark

4 $4{,}356 - 2{,}788 =$

1 mark

5 $\dfrac{7}{8} + \dfrac{5}{8} =$

1 mark

6 $3{,}926 \times 7 =$

1 mark

Test 18

7   25% of 1,000 =

1 mark

8 $368 \div 6 =$

1 mark

9 $7^2 \times 4 =$

1 mark

10 $763 \times 11 =$

2 marks

11

1	2	9	9	1	0

Show your method

2 marks

12 $2\frac{1}{3} + 1\frac{4}{5} =$

1 mark

10 min

1 $254 - 7 =$

1 mark

2 $\frac{2}{3}$ of $36 =$

1 mark

3 $1{,}000 \times 6 =$

1 mark

4 $245 \times 5 =$

1 mark

5 $45{,}678 - 26{,}824 =$

1 mark

6 $243.5 \div 10 =$

1 mark

7 $\frac{3}{4} + \frac{7}{8} =$

1 mark

8 Show your method

$$\begin{array}{r} 3\ 7\ 6\ 8 \\ \times \quad\quad 5\ 8 \\ \hline \end{array}$$

2 marks

9 $245{,}000 - 199{,}998 =$

1 mark

10 $100 - 3^2 \times 6 =$

1 mark

11 $\frac{5}{8}$ of $64 =$

1 mark

12 $\frac{4}{5} \div 6 =$

1 mark

10 min

1 $\dfrac{3}{10} + \dfrac{1}{10} =$

1 mark

2 $1,000 \times 9 =$

1 mark

3 $63 \div 9 =$

1 mark

4 $240 \div 6 =$

1 mark

5 $23,457 + 199 =$

1 mark

6 $700 \times 8 =$

1 mark

7 $4^3 \times 10 =$

1 mark

8 $\dfrac{4}{7} \times 8 =$

1 mark

9 $\dfrac{7}{100} \times 10 =$

1 mark

10

Show your method

| 1 | 8 | 4 | 6 | 1 | 0 |

2 marks

11 $\dfrac{3}{5} + \dfrac{2}{3} =$

1 mark

12 $7.37 \times 8 =$

1 mark

10 min

1   105 − 10 =

1 mark

2   12 × 8 =

1 mark

3   4,356 − 2,123 =

1 mark

4   0.5 + 0.75 =

1 mark

5   154 ÷ 10 =

1 mark

6   125 × 5 =

1 mark

7 $489 \times 6 =$

1 mark

8 $1,458 \div 9 =$

1 mark

9 $-36 - 54 =$

1 mark

10 $\dfrac{3}{4} - \dfrac{2}{7} =$

1 mark

11 $\dfrac{1}{9} \times \dfrac{3}{5} =$

1 mark

12 $23,567 \div 1,000 =$

1 mark

Test 21 total marks ..../12

10 min

1 $6 \times 4 = 2 \times \boxed{}$

1 mark

2 $48 \div 3 =$

1 mark

3 $24 \times 0 =$

1 mark

4 $\dfrac{2}{3} + \dfrac{2}{3} =$

1 mark

5 $100{,}000 \times 6 =$

1 mark

6 $4^2 \times 5 =$

1 mark

7 484 ÷ 7 =

1 mark

8 0.23 ÷ 10 =

1 mark

9

Show your method

4 2 | 6 9 3 0

2 marks

10 145 − 12 × 5 =

1 mark

11 45,900 + 35,100 =

1 mark

12 465 ÷ 12 =

1 mark

1 456 + 287 =

1 mark

2 89 = ☐ – 63

1 mark

3 $\frac{5}{6}$ of 48 =

1 mark

4 23,789 + 1,000 =

1 mark

5 12 × 5 × 4 =

1 mark

6 245,167 – 89,248 =

1 mark

7 $7,365 \div 1,000 =$

1 mark

8

$$\begin{array}{r} 2\ 5\ 6\ 7 \\ \times \qquad 2\ 8 \\ \hline \end{array}$$

Show your method

2 marks

9 $3,500 \div 5 =$

1 mark

10 $2\dfrac{3}{4} - \dfrac{7}{8} =$

1 mark

11 $\dfrac{3}{5} \times 11 =$

1 mark

12 15% of $480 =$

1 mark

10 min

1 $195 + 20 =$

1 mark

2 $144 \div 12 =$

1 mark

3 $\dfrac{3}{12} + \dfrac{5}{12} =$

1 mark

4 $4{,}001 - 2{,}789 =$

1 mark

5 $183 \div 3 =$

1 mark

6 $34{,}500 + 25{,}500 =$

1 mark

7 $1,155 \div 7 =$

1 mark

8 $5^3 \times 1^3 =$

1 mark

9

Show your method

3 6 | 5 6 1 6

2 marks

10 $3\frac{5}{7} + 2\frac{3}{6} =$

1 mark

11 $\frac{2}{5} \times \frac{1}{4} =$

1 mark

12 $\frac{5}{8} \div 5 =$

1 mark

1 $100 = \boxed{} \times 4$

1 mark

2 $9 \times \boxed{} = 63$

1 mark

3 $256 \times 3 =$

1 mark

4 $\dfrac{7}{10} + \dfrac{9}{10} =$

1 mark

5 $7 \div 100 =$

1 mark

6 $0.64 + 0.46 =$

1 mark

Test 25

7 15% of 140 =

1 mark

8 Show your method

4 8 | 3 1 2 0

2 marks

9 Show your method

9 5 6 2
×   3 5

2 marks

10 45,675 − 19,999 =

1 mark

11 8.69 × 7 =

1 mark

12 65% of 500 =

1 mark

Test 25 total marks ..../14

53

Well done

You have completed all the tests! Now check the answers and then write your scores in the score chart below.

Test	My Score
Test 1	/12
Test 2	/15
Test 3	/13
Test 4	/12
Test 5	/13
Test 6	/13
Test 7	/14
Test 8	/13
Test 9	/14
Test 10	/14
Test 11	/13
Test 12	/14
Test 13	/12
Test 14	/14
Test 15	/13
Test 16	/13
Test 17	/13
Test 18	/14
Test 19	/13
Test 20	/13
Test 21	/12
Test 22	/13
Test 23	/13
Test 24	/13
Test 25	/14
Total	/330

How did you do?

I did brilliantly!
Fabulous!

I did well.
Great stuff!

I did ok.
Well done – keep up the practice if you want to improve.

I didn't do so well.
Don't worry – there's still time to learn and practise. Why not try these tests again?

Answers

Question	Answer	Marks	Additional guidance
Test 1			
1	493	1	
2	150	1	
3	$\frac{4}{7}$	1	
4	6	1	
5	120	1	
6	363,366	1	
7	7	1	
8	21.8	1	
9	$\frac{1}{12}$	1	Accept equivalent fractions. Do not accept decimal answers.
10	100	1	
11	81	1	Do not accept 81%
12	$\frac{3}{8}$	1	Accept exact decimal equivalent, i.e. 0.375 Do not accept rounded or truncated decimals.
Test 2			
1	0	1	
2	60	1	
3	30	Up to 2	Award **TWO** marks for answer of 30 Working must be carried through to reach a final answer for **ONE** mark. Award **ONE** mark for a formal method of division with no more than ONE arithmetic error.
4	−18	1	
5	$\frac{9}{10}$	1	Accept exact decimal equivalent, i.e. 0.9

Question	Answer	Marks	Additional guidance
6	126	Up to 2	Award **TWO** marks for answer of 126 Working must be carried through to reach a final answer for **ONE** mark. Award **ONE** mark for a formal method of division with no more than ONE arithmetic error.
7	£9		Do not accept 9%
8	7,840	Up to 2	Award **TWO** marks for answer of 7,840 Award **ONE** mark for a formal method of long multiplication with no more than ONE arithmetic error. Working must be carried through to reach a final answer for **ONE** mark. Do not award any marks if the error is in the place value, e.g. the omission of the zero when multiplying by tens.
9	33.6	1	
10	$\frac{29}{40}$	1	Accept exact decimal equivalent, i.e. 0.725 Do not accept rounded or truncated decimals.
11	24.92	1	
12	75	1	
Test 3			
1	446	1	
2	6,609	1	
3	1,673	1	
4	41,342	1	
5	$\frac{13}{12}$ **OR** $1\frac{1}{12}$	1	Accept either the improper fraction or the mixed number.

Question	Answer	Marks	Additional guidance
6	0.012	1	Do not accept rounded or truncated decimals.
7	$\frac{2}{9}$	1	
8	150	Up to 2	Award **TWO** marks for answer of 150. Working must be carried through to reach a final answer for **ONE** mark. Award **ONE** mark for a formal method of division with no more than ONE arithmetic error.
9	$\frac{2}{15}$	1	Accept $33\frac{6}{10}$ and $33\frac{3}{5}$
10	211.5	1	
11	68.75	1	Accept 68 remainder 3 and $68\frac{3}{4}$
12	432	1	Do not accept 432%

Test 4

Question	Answer	Marks	Additional guidance
1	9	1	
2	28	1	
3	749	1	
4	9,050	1	
5	$\frac{6}{5}$ OR $1\frac{1}{5}$	1	Accept either the improper fraction or the mixed number. Also accept exact decimal equivalent, i.e. 1.2
6	0.27	1	
7	72	1	
8	$1\frac{3}{6}$ OR $1\frac{1}{2}$	1	Accept either answer. Also accept exact decimal equivalent, i.e. 1.5
9	$\frac{7}{16}$	1	
10	26	1	
11	$\frac{2}{20}$ OR $\frac{1}{10}$	1	Accept either answer. Also accept exact decimal equivalent, i.e. 0.1
12	600	1	Do not accept 600%

Test 5

Question	Answer	Marks	Additional guidance
1	446	1	
2	2,151	1	
3	$\frac{10}{10}$ OR 1	1	Accept either answer.
4	5	1	
5	1,203	1	
6	62,134	1	
7	20	1	Do not accept 20%
8	140,800	Up to 2	Award **TWO** marks for answer of 140,800. Award **ONE** mark for a formal method of long multiplication with no more than ONE arithmetic error. Working must be carried through to reach a final answer for **ONE** mark. Do not award any marks if the error is in the place value, e.g. the omission of the zero when multiplying by tens.
9	$\frac{43}{30}$ OR $1\frac{13}{30}$	1	Accept either the improper fraction or the mixed number.
10	$9\frac{3}{5}$	1	Accept the exact decimal equivalent, i.e. 9.6
11	$\frac{3}{20}$	1	Accept exact decimal equivalent, i.e. 0.15
12	2,750	1	

Test 6

Question	Answer	Marks	Additional guidance
1	100	1	
2	36	1	
3	180	1	
4	$\frac{11}{7}$ OR $1\frac{4}{7}$	1	Accept either the improper fraction or the mixed number.
5	61	1	
6	343	1	
7	900,000	1	
8	$3\frac{8}{9}$	1	

Question	Answer	Marks	Additional guidance
9	250	1	
10	120	1	Do not accept 120%
11	$\frac{3}{20}$	1	
12	56	Up to 2	Award **TWO** marks for answer of 56 Working must be carried through to reach a final answer for **ONE** mark. Award **ONE** mark for a formal method of division with no more than ONE arithmetic error.

Test 7

Question	Answer	Marks	Additional guidance
1	208	1	
2	$\frac{1}{16}$	1	
3	1,416	1	
4	12,775	1	
5	5,840	Up to 2	Award **TWO** marks for answer of 5,840 Award **ONE** mark for a formal method of long multiplication with no more than ONE arithmetic error. Working must be carried through to reach a final answer for **ONE** mark. Do not award any marks if the error is in the place value, e.g. the omission of the zero when multiplying by tens.
6	$\frac{12}{8}$ **OR** $1\frac{4}{8}$ **OR** $1\frac{1}{2}$	1	Accept equivalent mixed numbers, fractions or exact decimal equivalent, i.e. 1.5
7	241	1	
8	1,789	1	
9	45	1	
10	17.5 **OR** $17\frac{5}{10}$ **OR** $17\frac{1}{2}$	1	Accept any one of these answers. Do NOT accept answers with % at the end.

Question	Answer	Marks	Additional guidance
11	$\frac{2}{20}$ **OR** $\frac{1}{10}$	1	Accept either answer. Also accept exact decimal equivalent, i.e. 0.1
12	156	Up to 2	Award **TWO** marks for answer of 156 Working must be carried through to reach a final answer for **ONE** mark. Award **ONE** mark for a formal method of division with no more than ONE arithmetic error.

Test 8

Question	Answer	Marks	Additional guidance
1	260	1	
2	250	1	
3	72	1	
4	4,910	1	
5	$\frac{23}{15}$ **OR** $1\frac{8}{15}$	1	Accept either the improper fraction or the mixed number.
6	700	1	
7	10	1	Do not accept $6\frac{12}{3}$
8	40	1	
9	96	Up to 2	Award **TWO** marks for answer of 96 Working must be carried through to reach a final answer for **ONE** mark. Award **ONE** mark for a formal method of division with no more than ONE arithmetic error.
10	81	1	Do not accept 81%
11	$\frac{2}{6}$ **OR** $\frac{1}{3}$	1	Accept either answer.
12	44.7	1	

Test 9

Question	Answer	Marks	Additional guidance
1	77	1	
2	492	1	
3	1,000	1	
4	484	1	
5	810	1	
6	1.02	1	

Question	Answer	Marks	Additional guidance
7	455,938	Up to 2	Award **TWO** marks for answer of 455,938 Award **ONE** mark for a formal method of short or long multiplication with no more than ONE arithmetic error. Working must be carried through to reach a final answer for **ONE** mark. Do not award any marks if the error is in the place value, e.g. the omission of the zero when multiplying by tens.
8	10,458	1	
9	75%	1	
10	86	Up to 2	Award **TWO** marks for answer of 86 Working must be carried through to reach a final answer for **ONE** mark. Award **ONE** mark for a formal method of division with no more than ONE arithmetic error.
11	$\frac{3}{28}$	1	
12	$\frac{2}{30}$ **OR** $\frac{1}{15}$	1	Accept either answer.

Test 10

1	102	1	
2	6	1	
3	2,268	1	
4	26,892	1	
5	631	1	
6	$\frac{25}{8}$ **OR** $3\frac{1}{8}$	1	Accept either answer. Do NOT accept answers such as $1\frac{17}{8}$
7	22,800	1	

8	28	1	Do NOT accept 228 (100 − 24 = 76, 76 × 3 = 228)
9	6.25	1	Accept $6\frac{3}{12}$ and $6\frac{1}{4}$
10	89	Up to 2	Award **TWO** marks for answer of 89 Working must be carried through to reach a final answer for **ONE** mark. Award **ONE** mark for a formal method of division with no more than ONE arithmetic error.
11	107,847	Up to 2	Award **TWO** marks for answer of 107,847 Award **ONE** mark for a formal method of short or long multiplication with no more than ONE arithmetic error. Working must be carried through to reach a final answer for **ONE** mark. Do not award any marks if the error is in the place value, e.g. the omission of the zero when multiplying by tens.
12	256.4	1	

Test 11

1	868	1	
2	$\frac{6}{8}$ **OR** $\frac{3}{4}$	1	Accept either answer. Also accept exact decimal equivalent, i.e. 0.75
3	105	1	
4	40,000	1	
5	1,250	1	
6	0.034	1	Do not accept rounded or truncated decimals.
7	23%	1	
8	−28	1	

Question	Answer	Marks	Additional guidance
9	$\frac{2}{40}$ OR $\frac{1}{20}$	1	Accept either answer. Also accept exact decimal equivalent, i.e. 0.05
10	317.45	1	
11	84	1	Do not accept 84%
12	24	Up to 2	Award **TWO** marks for answer of 24 Working must be carried through to reach a final answer for **ONE** mark. Award **ONE** mark for a formal method of division with no more than ONE arithmetic error.
Test 12			
1	64	1	
2	140	1	
3	25	1	
4	6,000	1	
5	915	1	
6	1,948	1	
7	5,606,296	Up to 2	Award **TWO** marks for answer of 5,606,296 Award **ONE** mark for a formal method of long multiplication with no more than ONE arithmetic error. Working must be carried through to reach a final answer for **ONE** mark. Do not award any marks if the error is in the place value, e.g. the omission of the zero when multiplying by tens.
8	$2\frac{5}{8}$	1	
9	$\frac{15}{40}$ OR $\frac{3}{8}$	1	Accept either answer. Also accept exact decimal equivalent, i.e. 0.375 Do not accept rounded or truncated decimals.

Question	Answer	Marks	Additional guidance
10	0.225	1	Do not accept rounded or truncated decimals.
11	60%	1	
12	45 remainder 1 OR $45\frac{1}{11}$	Up to 2	Award **TWO** marks for answer of 45 remainder 1 **OR** $45\frac{1}{11}$. Working must be carried through to reach a final answer for **ONE** mark. Award **ONE** mark for a formal method of division with no more than ONE arithmetic error. Do not award mark if remainder is excluded.
Test 13			
1	0.25	1	Accept $\frac{1}{4}$
2	403	1	
3	435	1	
4	80	1	
5	7,079	1	
6	36	1	
7	7,500	1	
8	$\frac{9}{50}$	1	Accept exact decimal equivalent, i.e. 0.18
9	125	1	
10	82 remainder 1 OR $82\frac{1}{6}$	1	
11	28	1	$45 \div 9$ must be carried out first and then added to 23
12	70.08	1	
Test 14			
1	0.45	1	
2	$\frac{7}{5}$ OR $1\frac{2}{5}$	1	Accept either the improper fraction or the mixed number. Also accept exact decimal equivalent, i.e. 1.4
3	51	1	

Question	Answer	Marks	Additional guidance
4	4	1	
5	56	1	
6	0	1	
7	10,783	1	
8	65,835	Up to 2	Award **TWO** marks for answer of 65,835 Award **ONE** mark for a formal method of long multiplication with no more than ONE arithmetic error. Working must be carried through to reach a final answer for **ONE** mark. Do not award any marks if the error is in the place value, e.g. the omission of the zero when multiplying by tens.
9	168	1	Do not accept 168%
10	84	1	
11	458	Up to 2	Award **TWO** marks for answer of 458. Working must be carried through to reach a final answer for **ONE** mark. Award **ONE** mark for a formal method of division with no more than ONE arithmetic error.
12	49	1	Do not accept 49%
Test 15			
1	$\frac{2}{8}$ **OR** $\frac{1}{4}$	1	Accept either answer. Also accept exact decimal equivalent, i.e. 0.25
2	262	1	
3	259	1	
4	240	1	
5	400,000	1	
6	35,567	1	

Question	Answer	Marks	Additional guidance
7	$\frac{6}{12}$ **OR** $\frac{3}{6}$ **OR** $\frac{1}{2}$	1	Accept any one of these answers. Also accept exact decimal equivalent, i.e. 0.5
8	$\frac{42}{8}$ **OR** $5\frac{2}{8}$ **OR** $5\frac{1}{4}$	1	Accept any one of these answers. Also accept exact decimal equivalent, i.e. 5.25
9	69,525	Up to 2	Award **TWO** marks for answer of 69,525 Award **ONE** mark for a formal method of short or long multiplication with no more than ONE arithmetic error. Working must be carried through to reach a final answer for **ONE** mark. Do not award any marks if the error is in the place value, e.g. the omission of the zero when multiplying by tens.
10	$4\frac{3}{40}$ **OR** $4\frac{6}{80}$	1	Do NOT accept $3\frac{43}{40}$
11	21.2	1	
12	$\frac{6}{12}$ **OR** $\frac{3}{6}$ **OR** $\frac{1}{2}$	1	Accept any one of these answers. Also accept exact decimal equivalent, i.e. 0.5
Test 16			
1	300	1	
2	217	1	
3	14	1	
4	9,462	1	
5	3,822	1	
6	$\frac{3}{8}$	1	
7	182,400	1	
8	0.0385	1	Do not accept rounded or truncated decimals.
9	57	1	

Question	Answer	Marks	Additional guidance
10	322,020	Up to 2	Award **TWO** marks for answer of 322,020. Award **ONE** mark for a formal method of short or long multiplication with no more than ONE arithmetic error. Working must be carried through to reach a final answer for **ONE** mark. Do not award any marks if the error is in the place value, e.g. the omission of the zero when multiplying by tens.
11	$\frac{7}{36}$	1	
12	270	1	Do not accept 270%

Test 17

Question	Answer	Marks	Additional guidance
1	127	1	
2	8	1	
3	175	1	
4	63	1	
5	9,090	1	
6	125	1	
7	120	1	
8	85	Up to 2	Award **TWO** marks for answer of 85. Working must be carried through to reach a final answer for **ONE** mark. Award **ONE** mark for a formal method of division with no more than ONE arithmetic error.
9	35	1	
10	$\frac{6}{20}$ OR $\frac{3}{10}$	1	Accept either answer. Also accept exact decimal equivalent, i.e. 0.3
11	1,234	1	
12	156.8 OR 156 remainder 4 OR $156\frac{4}{5}$	1	Accept any one of these answers.

Test 18

Question	Answer	Marks	Additional guidance
1	95	1	
2	310	1	
3	81	1	
4	1,568	1	
5	$\frac{12}{8}$ OR $1\frac{4}{8}$ OR $1\frac{1}{2}$ OR $1\frac{2}{4}$	1	Accept any one of these answers. Also accept exact decimal equivalent, i.e. 1.5
6	27,482	1	
7	250	1	
8	61 remainder 2 OR $61\frac{2}{6}$ OR $61\frac{1}{3}$	1	Accept any one of these answers.
9	196	1	
10	8,393	Up to 2	Award **TWO** marks for answer of 8,393. Award **ONE** mark for any method with no more than ONE arithmetic error. Working must be carried through to reach a final answer for **ONE** mark. Do not award any marks if the error is in the place value, e.g. the omission of the zero when multiplying by tens.
11	825 remainder 10 OR $825\frac{10}{12}$ OR $825\frac{5}{6}$	Up to 2	Award **TWO** marks for any one of these answers. Working must be carried through to reach a final answer for **ONE** mark. Award **ONE** mark for a formal method of division with no more than ONE arithmetic error.

Question	Answer	Marks	Additional guidance
12	$4\frac{2}{15}$	1	Do not accept $3\frac{17}{15}$
Test 19			
1	247	1	
2	24	1	
3	6,000	1	
4	1,225	1	
5	18,854	1	
6	24.35	1	
7	$\frac{13}{8}$ **OR** $1\frac{5}{8}$	1	Accept either the improper fraction or the mixed number. Also accept exact decimal equivalent, i.e. 1.625 Do not accept rounded or truncated decimals.
8	218,544	Up to 2	Award **TWO** marks for answer of 218,544 Award **ONE** mark for a formal method of long multiplication with no more than ONE arithmetic error. Working must be carried through to reach a final answer for **ONE** mark. Do not award any marks if the error is in the place value, e.g. the omission of the zero when multiplying by tens.
9	45,002	1	
10	46	1	
11	40	1	
12	$\frac{4}{30}$ **OR** $\frac{2}{15}$	1	Accept either answer.
Test 20			
1	$\frac{4}{10}$ **OR** $\frac{2}{5}$	1	Accept either answer. Also accept exact decimal equivalent, i.e. 0.4
2	9,000	1	
3	7	1	
4	40	1	
5	23,656	1	

Question	Answer	Marks	Additional guidance
6	5,600	1	
7	640	1	
8	$\frac{32}{7}$ **OR** $4\frac{4}{7}$	1	Accept either the improper fraction or the mixed number.
9	$\frac{7}{10}$ **OR** $\frac{70}{100}$	1	Accept exact decimal equivalent, i.e. 0.7
10	256 remainder 2 **OR** $256\frac{2}{18}$ **OR** $256\frac{1}{9}$ **OR** 256.11	Up to 2	Award **TWO** marks for any one of these answers. Working must be carried through to reach a final answer for **ONE** mark. Award **ONE** mark for a formal method of division with no more than ONE arithmetic error.
11	$\frac{19}{15}$ **OR** $1\frac{4}{15}$	1	Accept either the improper fraction or the mixed number.
12	58.96	1	
Test 21			
1	95	1	
2	96	1	
3	2,233	1	
4	1.25	1	Also accept $1\frac{1}{4}$
5	15.4	1	Also accept $15\frac{4}{10}$ **OR** $15\frac{2}{5}$
6	625	1	
7	2,934	1	
8	162	1	
9	−90	1	
10	$\frac{13}{28}$	1	
11	$\frac{3}{45}$ **OR** $\frac{1}{15}$	1	Accept either answer.
12	23.567	1	Do not accept rounded or truncated decimals.
Test 22			
1	12	1	
2	16	1	
3	0	1	

Question	Answer	Marks	Additional guidance
4	$\frac{4}{3}$ OR $1\frac{1}{3}$	1	Accept either the improper fraction or the mixed number.
5	600,000	1	
6	80	1	
7	69 remainder 1 OR $69\frac{1}{7}$	1	Accept either answer.
8	0.023	1	Accept $\frac{23}{1000}$
9	165	Up to 2	Award **TWO** marks for answer of 165. Working must be carried through to reach a final answer for **ONE** mark. Award **ONE** mark for a formal method of division with no more than ONE arithmetic error.
10	85	1	
11	81,000	1	
12	38.75 OR 38 remainder 9	1	Also accept $38\frac{9}{12}$ OR $38\frac{3}{4}$

Test 23

Question	Answer	Marks	Additional guidance
1	743	1	
2	152	1	
3	40	1	
4	24,789	1	
5	240	1	
6	155,919	1	
7	7.365	1	Do not accept rounded or truncated decimals.

Question	Answer	Marks	Additional guidance
8	71,876	Up to 2	Award **TWO** marks for answer of 71,876. Award **ONE** mark for a formal method of long multiplication with no more than ONE arithmetic error. Working must be carried through to reach a final answer for **ONE** mark. Do not award any marks if the error is in the place value, e.g. the omission of the zero when multiplying by tens.
9	700	1	
10	$1\frac{7}{8}$ OR $\frac{15}{8}$	1	Accept either the improper fraction or the mixed number. Also accept exact decimal equivalent, i.e. 1.875
11	$\frac{33}{5}$ OR $6\frac{3}{5}$	1	Accept either the improper fraction or the mixed number. Also accept exact decimal equivalent, i.e. 6.6
12	72	1	Do not accept 72%

Test 24

Question	Answer	Marks	Additional guidance
1	215	1	
2	12	1	
3	$\frac{8}{12}$ OR $\frac{4}{6}$ OR $\frac{2}{3}$	1	Accept any one of these answers.
4	1,212	1	
5	61	1	
6	60,000	1	
7	165	1	
8	125	1	

Question	Answer	Marks	Additional guidance
9	156	Up to 2	Award **TWO** marks for the answer of 156 Working must be carried through to reach a final answer for **ONE** mark. Award **ONE** mark for a formal method of division with no more than ONE arithmetic error.
10	$6\frac{9}{42}$ **OR** $6\frac{3}{14}$	1	Accept either answer. Do not accept $5\frac{51}{42}$
11	$\frac{2}{20}$ **OR** $\frac{1}{10}$	1	Accept either answer. Also accept exact decimal equivalent, i.e. 0.1
12	$\frac{5}{40}$ **OR** $\frac{1}{8}$	1	Accept either answer. Also accept exact decimal equivalent, i.e. 0.125 Do not accept rounded or truncated decimals.

Test 25

Question	Answer	Marks	Additional guidance
1	25	1	
2	7	1	
3	768	1	
4	$\frac{16}{10}$ **OR** $\frac{8}{5}$ $1\frac{6}{10}$ **OR** $1\frac{3}{5}$	1	Accept any one of these answers. Also accept exact decimal equivalent, i.e. 1.6

5	0.07	1	
6	1.1	1	Accept $1\frac{1}{10}$ Do not accept 1.10
7	21	1	Do not accept 21%
8	65	Up to 2	Award **TWO** marks for the answer of 65 Working must be carried through to reach a final answer for **ONE** mark. Award **ONE** mark for a formal method of division with no more than ONE arithmetic error.
9	334,670	Up to 2	Award **TWO** marks for answer of 334,670 Award **ONE** mark for a formal method of long multiplication with no more than ONE arithmetic error. Working must be carried through to reach a final answer for **ONE** mark. Do not award any marks if the error is in the place value, e.g. the omission of the zero when multiplying by tens.
10	25,676	1	
11	60.83	1	
12	325	1	Do not accept 325%